Fundamentals of COLLEGE ASTRONOMY EXPERIMENTS

ONLINE

MICHAEL C. LOPRESTO
STEVEN MURRELL

Kendall Hunt
publishing company

Cover images courtesy of NASA.

Kendall Hunt
publishing company

www.kendallhunt.com
Send all inquiries to:
4050 Westmark Drive
Dubuque, IA 52004-1840

Copyright © 2017 by Michael C. LoPresto and Steven Murrell

PAK ISBN: 978-1-5249-4892-4
Text Alone ISBN: 978-1-5249-4893-1

Kendall Hunt Publishing Company has the exclusive rights to reproduce this work, to prepare derivative works from this work, to publicly distribute this work, to publicly perform this work and to publicly display this work.

All rights reserved. No part of this publication may be reproduced, stored in a retrieval system, or transmitted, in any form or by any means, electronic, mechanical, photocopying, recording, or otherwise, without the prior written permission of the copyright owner.

Published in the United States of America

CONTENTS

1 The Scale of the Universe. 1

2 The Sun and the Seasons. 7

3 The Phases of the Moon 11

4 Kepler's Third Law of Planetary Motion 17

5 Classifying Planets 23

6 The Transit Method of Extra Solar Planet Detection 27

7 The Radial Velocity Method of Extra Solar Planet Detection . . . 31

8 Terrestrial Planet Surfaces 35

9 Telescopes and Atmospheric Absorption 43

10 Blackbody Curves 47

11 The Hertzsprung–Russel Diagram 51

12 Hubble's Law 55

13 Epilogue—The Drake Equation 59

EXPERIMENT #1

The Scale of the Universe

The sizes of astronomical objects compared to the distances between them vary when compared at different scales. In this activity you will determine the distance/size ratio for each of the four basic scales of the universe and compare them.

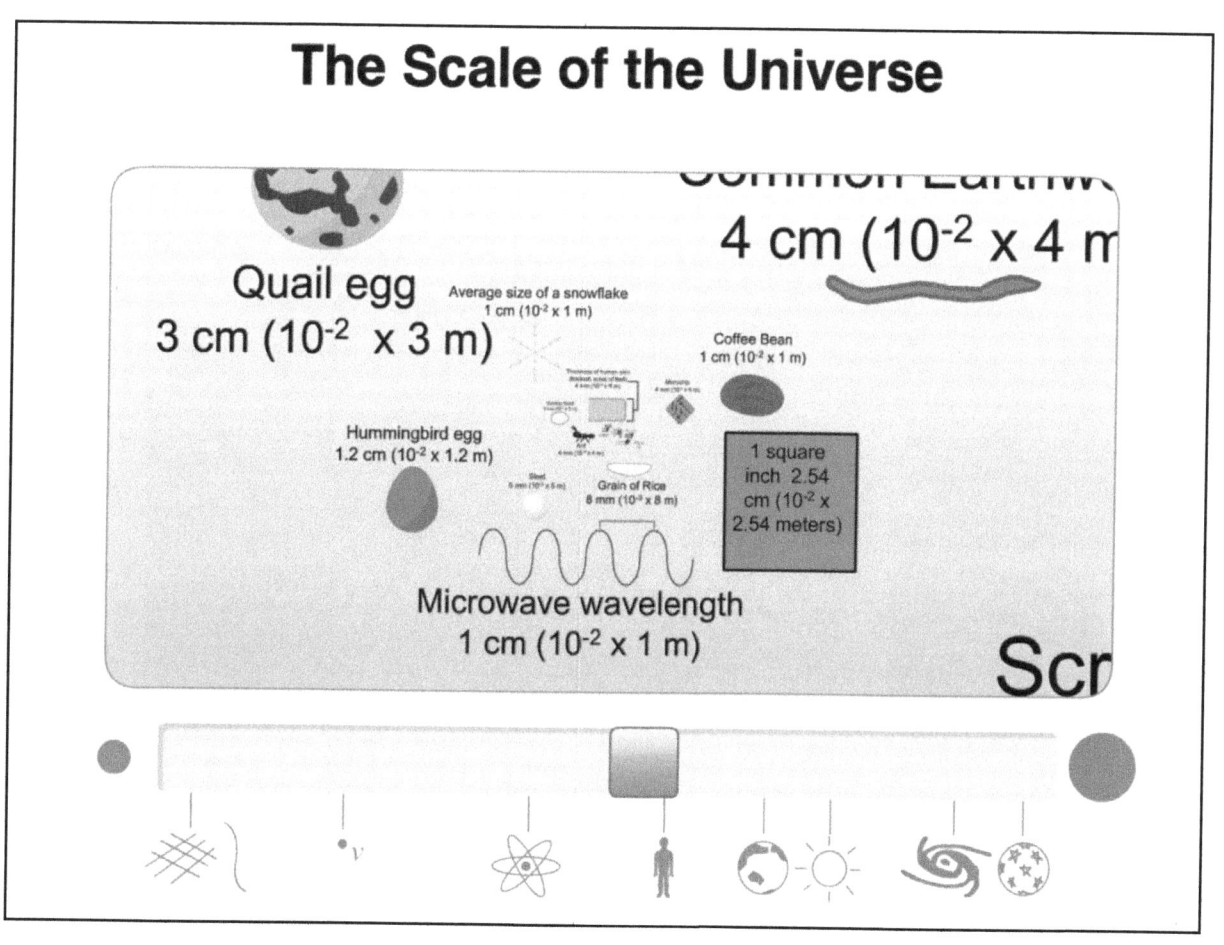

FIGURE 1.1 The Scale of the Universe Tool
Copyright © 2017 by CARY & MICHAEL HUANG. Reprinted by permission.

1. Earth–Moon

   Use the *Scale of the Universe Tool* shown in **Figure 1.1** to find the diameter of Earth and Distance from Earth to the Moon.

   Diameter of Earth = _____ km = SIZE

   Earth–Moon Distance = _____ km = DISTANCE

   Now divide the DISTANCE by the SIZE;

   DISTANCE/SIZE = _____ / _____ =

   The *order of magnitude* of a number is the number rounded off to the nearest 1, 10, 100, or 1000, etc., depending on which it is closest to.

   The order of magnitude of the distance/size ratio for the Earth–Moon system is (circle one below);

   1   10   100   1000   10,000   100,000   1,000,000   10,000,000   100,000,000

2. Sun–Earth

Use the *Scale of the Universe Tool* shown in **Figure 1.1** to find the diameter of Earth and Distance from Earth to the Sun.

Diameter of Earth = _____ km = SIZE

Earth–Sun Distance = _____ km = DISTANCE

DISTANCE/SIZE = _____ / _____ =

The order of magnitude of the distance/size ratio for the Sun–Earth system is (circle one below);

1   10   100   1000   10,000   100,000   1,000,000   10,000,000   100,000,000

The order of magnitude of the distance to size ratio of the Sun–Earth system is _____ / _____ = _____ times that of the Earth–Moon system.

3. Sun–Nearest Other Star

   Use the *Scale of the Universe Tool* shown in **Figure 1.1** to find the diameter of the Sun and Distance from the Sun to the nearest other star, Proxima Centauri.

   Diameter of Sun = _____ km = SIZE

   Distance to Proxima Centauri = _____ km = DISTANCE

   DISTANCE/SIZE = _____ / _____ =

   The order of magnitude of the distance/size ratio for the Sun–Nearest other Star is (circle one below);

   1   10   100   1000   10,000   100,000   1,000,000   10,000,000   100,000,000

   The order of magnitude of the distance to size ratio of the Sun–Nearest Star system is _____ / _____ = _____ times that of the Sun–Earth system.

4. Milky Way Galaxy–Andromeda Galaxy

Use the *Scale of the Universe Tool* shown in **Figure 1.1** to find the diameter of our Milky Way Galaxy. The Distance from the Milky Way Galaxy to the nearest other galaxy, the Andromeda Galaxy, is 3,000,000 light years.

Diameter of Milky Way Galaxy = _____ light years = SIZE

Distance to Andromeda Galaxy = 3,000,000 light years = DISTANCE

DISTANCE/SIZE = _____ / _____ =

The order of magnitude of the distance/size ratio for the Milky Way–Andromeda system is (circle one below);

1   10   100   1000   10,000   100,000   1,000,000   10,000,000   100,000,000

The order of magnitude of the distance/size ratio Milky Way–Andromeda system is most similar to that of which other system you have examined? (circle one)

Earth–Moon   Earth–Sun   Sun–Nearest Star

In which system(s) is the DISTANCE/SIZE ratio of the <u>largest</u> order of magnitude?

In which system(s) is the DISTANCE/SIZE ratio of the <u>smallest</u> order of magnitude?

In which two systems are the DISTANCE/SIZE ratios of the most <u>similar</u> order of magnitude?

Based on the distance/size ratios for each system, in which system are the objects the most isolated from one another?

> Now that you have finished the experiment,
> use your answers to take the online quiz on the
> *Fundamental College Astronomy Experiments* **ONLINE** website.

EXPERIMENT #2

The Sun and the Seasons

The Earth's rotational axis is tilted with respect to its orbital path around the Sun. This is the cause of the seasonal variations we experience. In this experiment you will observe the changes that occur in the Sun's altitude and the hours of daylight and darkness that occur at different times of the year from different locations on Earth.

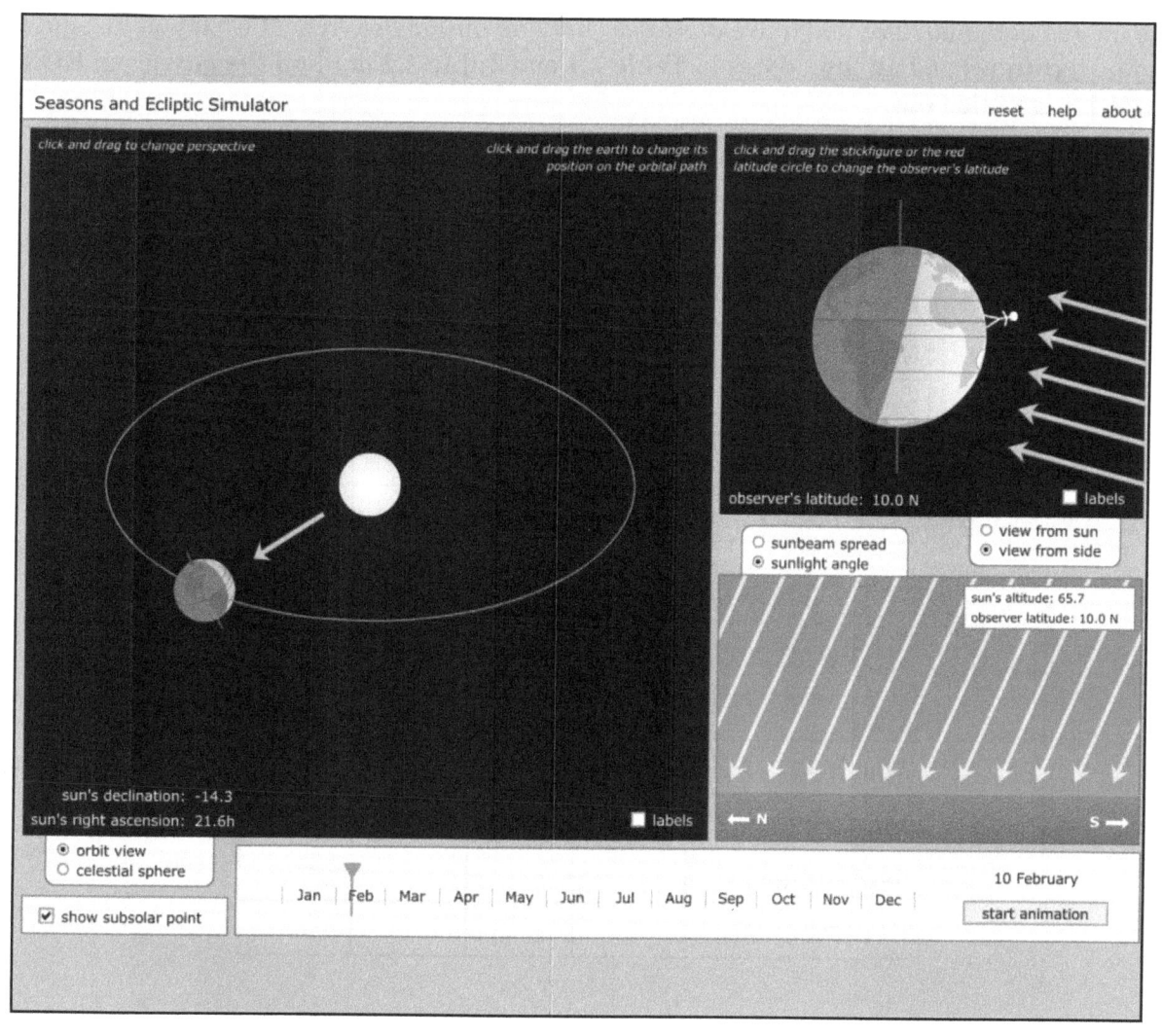

FIGURE 2.1 The Seasons and Ecliptic Simulator
Simulation downloaded from the Astronomy Education at the University of Nebraska-Lincoln Website (http://astro.unl.edu).

You have control of *two* things on the *Season's and Ecliptic Simulator*;

▶ You can drag the **red date indicator** to any month of the year.

▶ You can move the **observer standing on Earth** to any latitude.

Move the *red date indicator* to about March 21 and move the *observer standing on Earth* to an observer's *latitude* of about 40°N. Now record the *Sun's Altitude* on March 21 in **Table 2.1** below.

The red line at the observer's latitude can be used to estimate the *Hours the Sun is Up* by estimating what percentage of the 24-hour day the observer will be in daylight and in darkness as Earth rotates. Put another way, what fraction of the red line is in daylight and what fraction is in darkness? Take the daylight fraction and multiply by 24. This gives the estimated number of hours of daylight for that date and latitude. Estimate the *Hours the Sun is Up* on March 21 from 40°N and record in **Table 2.2** below.

Use the *red date indicator* and move *observer standing on Earth* to repeat what you just did for **all four latitudes** and **all four dates** in Table 2.1 and Table 2.2 until **all the entries** in BOTH TABLES are filled out.

TABLE 2.1 Sun's Altitude

Date	40°N	70°N	20°N	40°S
March 21				
June 21				
September 21				
December 21				

TABLE 2.2 Hours the Sun is Up (0–24)

Date	40°N	70°N	20°N	40°S
March 21				
June 21				
September 21				
December 21				

> Answer the following questions about the Sun based on the data you filled out in **TABLES 2.1 and 2.2.**

At 40°N *latitude*, on which date did the Sun reach its highest *altitude*? How did the *Hours the Sun was Up* on that date compare to other times of year?

At 40°N *latitude*, on which date did the Sun reach its lowest *altitude*? How did the *Hours the Sun was Up* on that date compare to other times of year?

At 40°S *latitude*, on which date did the Sun reach its highest *altitude*? How did the *Hours the Sun was Up* on that date compare to other times of year?

At 40°S *latitude*, on which date did the Sun reach its lowest *altitude*? How did the *Hours the Sun was Up* on that date compare to other times of year?

On what date(s) was the *Hours the Sun was Up* the same everywhere on Earth?

Which *latitude(s)* has the greatest variation in *Hours the Sun is Up* throughout the year?

Which *latitude(s)* has the least variation in *Hours the Sun is Up* throughout the year?

Was there a *latitude* and date where the Sun was, or close to, directly overhead (90° altitude)? The point directly overhead is called the *zenith*. If so, at which latitude and date?

Did you ever observe a *latitude* and date where the Sun was up all night or close to it? This is called *Midnight Sun*. If so, at which latitude and what date?

Did you ever observe *latitude* and date where the Sun was up for little or no time at all? If so, at which latitude and what date?

> Now that you have finished the experiment,
> use your answers to take the online quiz on the
> Fundamental College Astronomy Experiments **ONLINE** website.

EXPERIMENT #3

The Phases of the Moon

As the Moon orbits Earth, the amount of the side of the Moon receiving light from the Sun that can be seen by an observer on Earth changes, changing the appearance of the Moon. These different appearances are called Lunar Phases. In this experiment you will observe the changing phases of the Moon and times that the Moon is visible as the Moon orbits Earth.

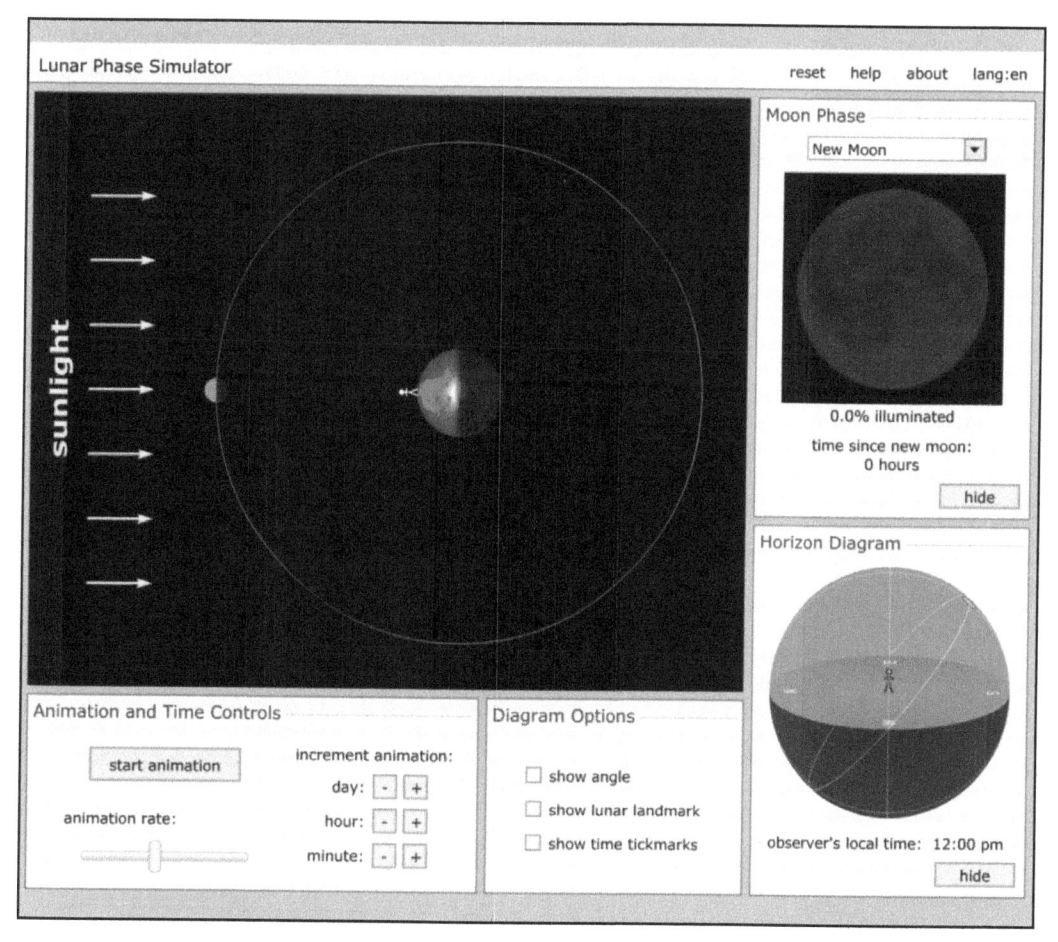

FIGURE 3.1 The Lunar Phase Simulator

Simulation downloaded from the Astronomy Education at the University of Nebraska-Lincoln Website (http://astro.unl.edu).

The main window of the *Lunar Phases Simulator* is a view from above Earth's North Pole of the Moon's orbit around Earth.

You have control of *two* things;

> ▶ You can drag the *moon* **around its orbit** to change its phase.
>
> ▶ You can move the *observer standing on Earth* to observe the moon from any location.

The window in the upper–right of the *Lunar Phases Simulator* display, Moon Phase, shows the appearance of the Moon to an **observer standing on Earth**. It should be set to *New Moon*, to begin. If it is not, press reset on the bar above it. In **Table 3.1** on the following page record the *% illuminated* and color in how much of the Moon appears dark. Note, *"% illuminated"* refers to the portion of the near side or "face" of the moon that is receiving sunlight.

Next, drag the **observer standing on Earth** to rotate Earth counter-clockwise while observing the *observer's local time* in the Horizon Diagram on the lower–right of the display. Record in **Table 3.1** the observer's local time that the Moon rises in the East, is high overhead in the South (the *transit* time), and sets in the West.

Now drag the moon in its orbit again counter-clockwise in its orbit to the three other Moon Phases given in **Table 3.1** and repeat the above procedure to record the same information for each of the other Moon Phases.

TABLE 3.1

Moon Phase	Appearance (color in part of Moon that appears dark)	% Illuminated	Rise Time	Transit Time	Set Time
New Moon					
First Quarter					
Full Moon					
Third Quarter					

Photos courtesy of NASA

> Use the *Lunar Phases Simulator* and the data you entered in **Table 3.1**, to answer the following questions:

During which Moon Phase do you see the entire face of the Moon?

During which Moon Phase(s) do you see half of the face of the Moon?

During which Moon Phase(s) do you see most of the face of the Moon but not all of it?

During which Moon Phase(s) do you see less than half of the face of the Moon but at least some of it?

During which Moon Phase do you not see any of the face of the Moon?

During which Moon Phase do you see the Moon mostly at night?

During which Moon Phase do you see the Moon first at night, then during the day?

During which Moon Phase do you see the Moon first during the day, then at night?

During which Moon Phase is the Moon up all day?

During which Moon Phase will you not see the Moon at all?

Now that you have finished the experiment,
use your answers to take the online quiz on the
Fundamental College Astronomy Experiments **ONLINE** website.

EXPERIMENT #4

Kepler's Third Law of Planetary Motion

Kepler's Third Law of Planetary Motion is a numerical relationship between a planet's orbital radius, its average distance from the Sun or *semimajor axis*, a, and its *orbital period*, p, the amount of time a planet takes to complete one orbit around the Sun.

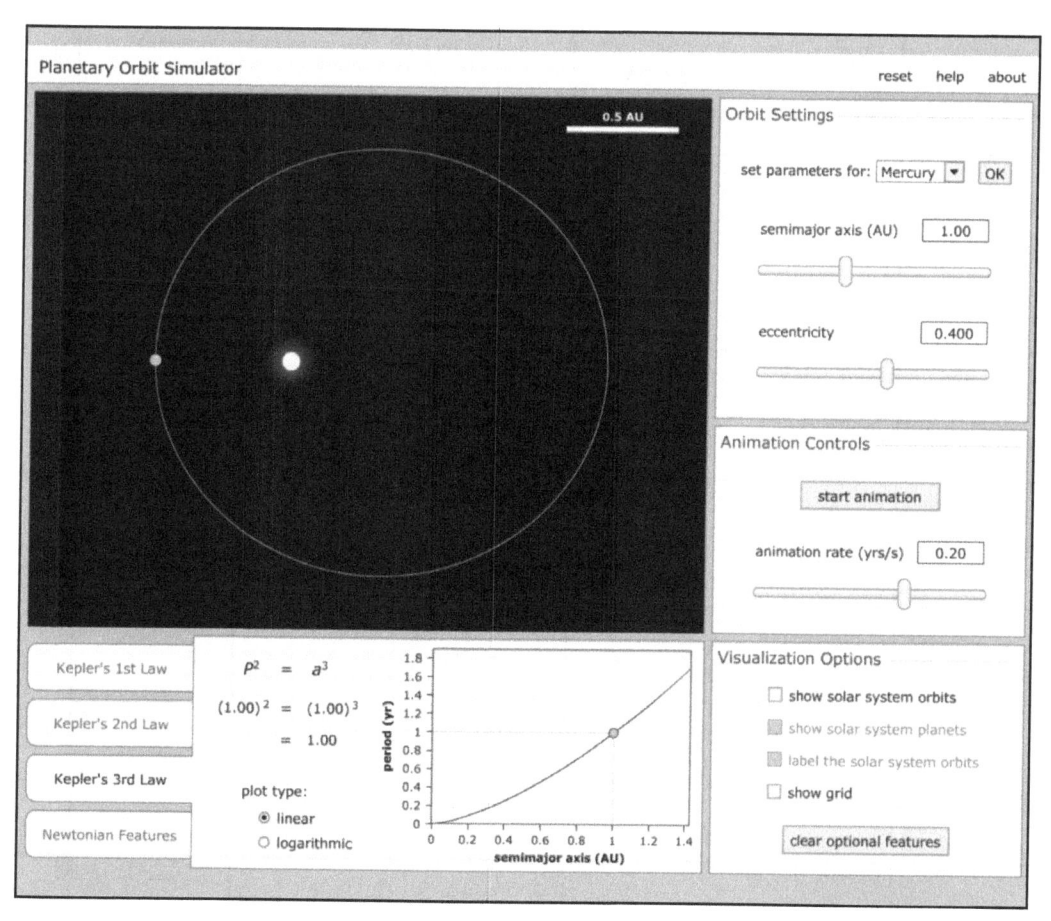

FIGURE 4.1 The **Planetary Orbit Simulator** set for Kepler's Third Law
Simulation downloaded from the Astronomy Education at the University of Nebraska-Lincoln Website (http://astro.unl.edu).

Select the button on *Planetary Orbit Simulator* for Kepler's Third Law as shown in **Figure 4.1**. Press reset on the bar above Orbit Settings on the upper-right to make sure the simulator is set for *Mercury*. Click "OK" next to Mercury to change the display to *Mercury's* orbit. The values for the *period*, p, and the *semimajor axis*, a, in the graph window below the display will also change to the values for *Mercury*. Record the values for the *period*, p, and the *semimajor axis*, a, for Mercury's orbit in **Table 4.1** below. Repeat this for each object listed in **Table 4.1**.

TABLE 4.1 Planetary Orbital Data

Object	Orbital Period, p (years)	Semimajor Axis, a (AU)	p^2	a^3
Mercury				
Venus				
Earth				
Mars				
Jupiter				
Saturn				
Uranus				
Neptune				
Pluto				

On the axis provided below, plot the *orbital periods,* p, vs. *the semimajor axis,* a, <u>for the planets *Mercury, Venus, Mars, Jupiter,* and *Saturn* ONLY</u>. These are the planets that can be seen with the unaided eye (those included in the data that Kepler himself used). The point for Earth is already plotted on the graph.

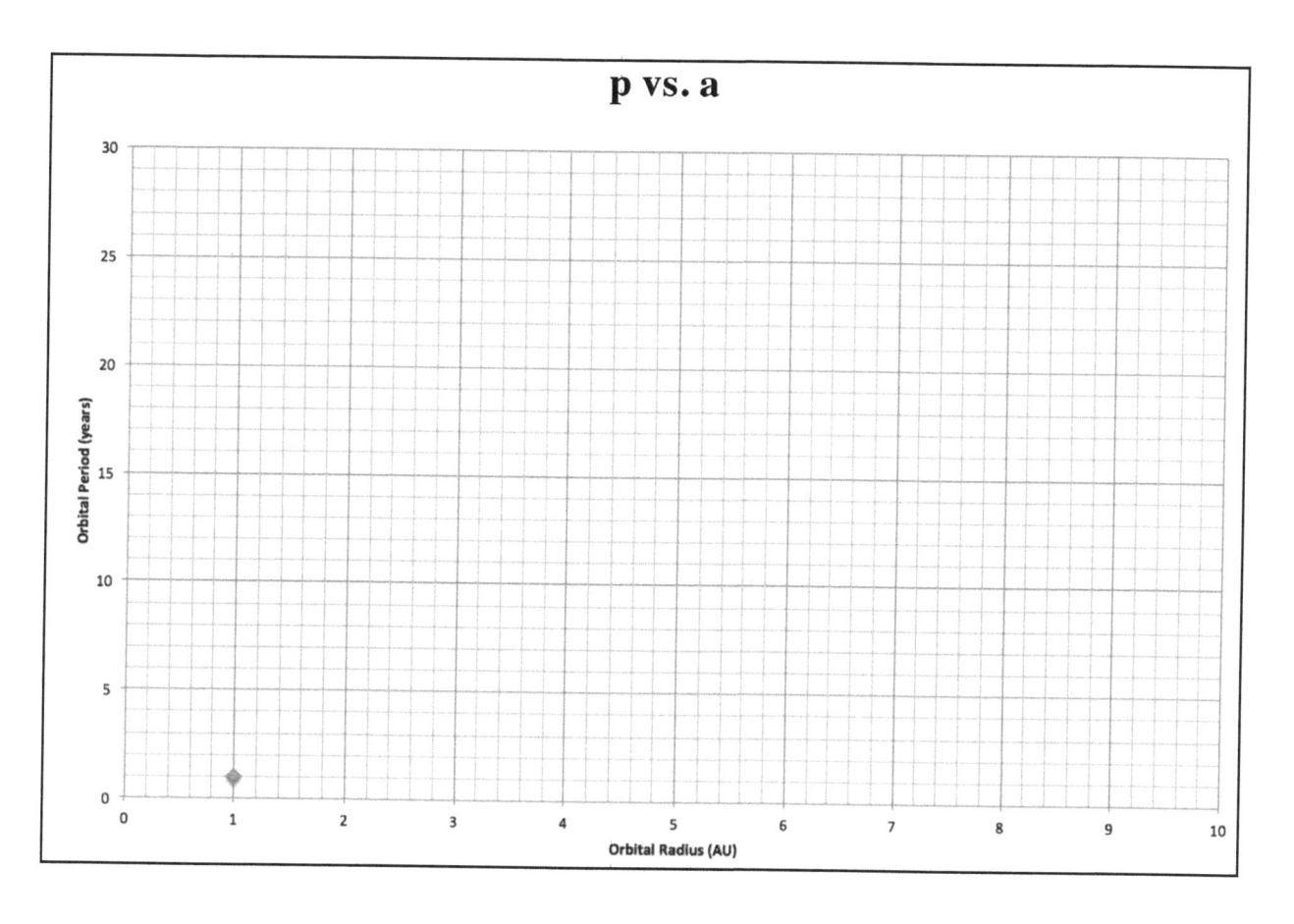

p vs. a

[Graph with x-axis "Orbital Radius (AU)" ranging 0 to 10 and y-axis "Orbital Period (years)" ranging 0 to 30, with Earth's point plotted at (1, 1).]

Extract

Now use your graph, or the graph window in the *Planetary Orbit Simulator,* to determine the orbital period, p, of an asteroid that has a *semimajor axis* of a = 4 AU. You can do this by setting the *semimajor axis* in the <u>Orbit Settings</u> window to 4 AU (or as close to 4 AU as you can).

$$a = 4 \text{ AU} \qquad p = \underline{\qquad} \text{ years}$$

Also determine *semimajor axis,* a, of another asteroid that has an *orbital period* of p = 4 years.

$$p = 4 \text{ years} \qquad a = \underline{\qquad} \text{ AU}$$

Now complete the last two columns of **TABLE 4.1** on the previous page. Squaring a number, p^2 means multiplying it by itself one more time, $p^2 = p \times p$. Cubing a number, a^3, means multiplying it by itself two more times, $a^3 = a \times a \times a$. Round off your answers for the inner planets, *Mercury* through *Mars* to two or three decimal points. Round off your answers for the outer planets, *Jupiter* and beyond, to the nearest whole number.

Now plot your data from the last two columns of **TABLE 4.1,** for the planets *Jupiter* and beyond ONLY, on the axis below. The point for Jupiter is already plotted on the graph.

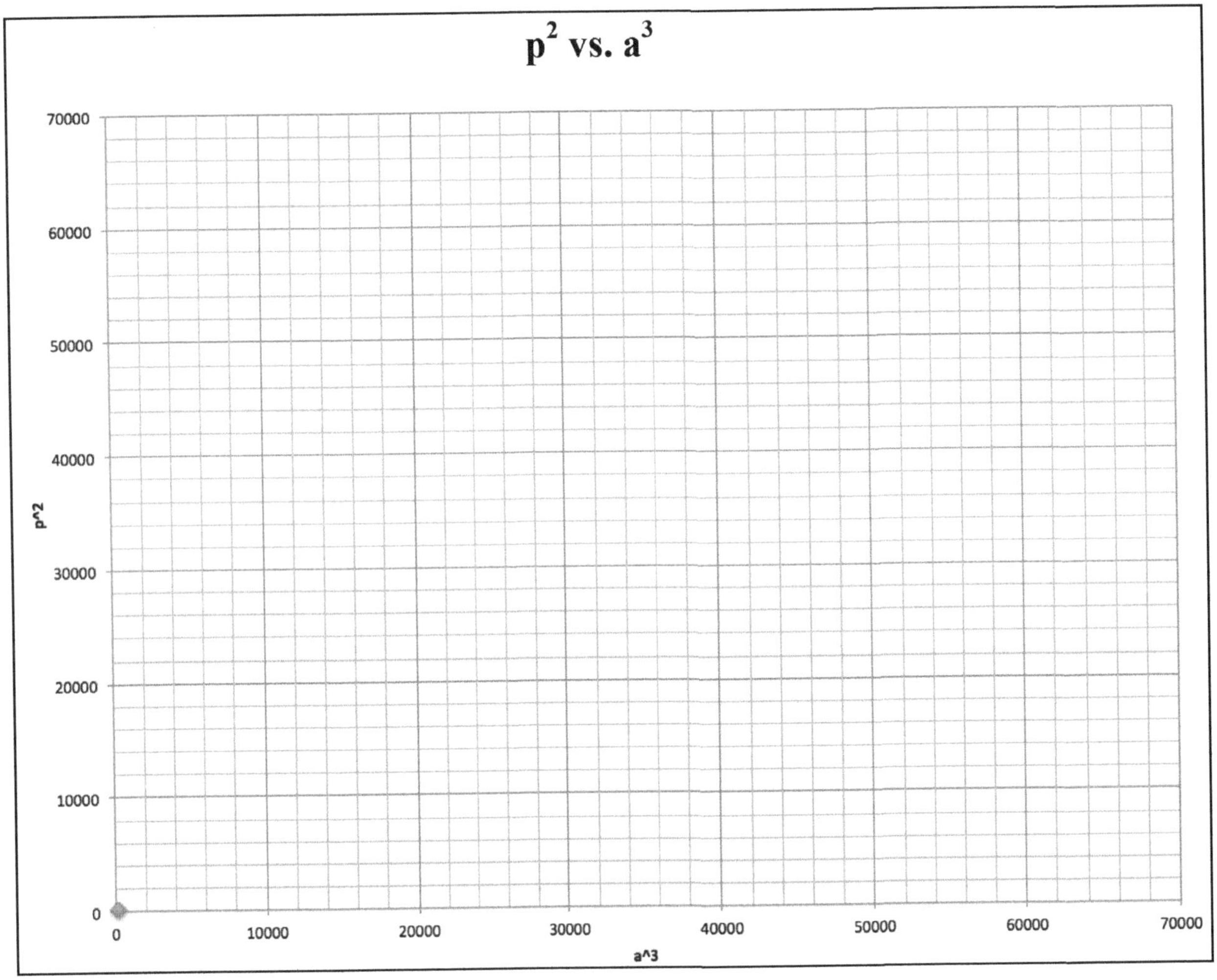

Extract

What is the shape of your graph?

Based on the shape of your graph and the values you calculated for **TABLE 4.1**, write the equation for *Kepler's Third Law of Planetary Motion*, the equation for the relationship between the *orbital period* and *semimajor axis* of an object in orbit around the Sun.

> Use your data in **TABLE 4.1** and/or your graphs and the *Planetary Orbit Simulator* to answer the questions below.

According to your graph of p vs. a, how do the *orbital periods* of planets further from the Sun compare to the orbital periods or planets that are closer to the Sun?

What is the shape of your graph of p vs. a?

What is the shape of your graph of p^2 vs. a^3?

Use your graph of p vs. a or the *Planetary Orbit Simulator* to determine the approximate *orbital period*, p, of an asteroid with an orbit of *semimajor axis* a = 3 AU.

Use your graph of p vs. a or the *Planetary Orbit Simulator* to determine the approximate *semimajor axis*, a, of an asteroid with an *orbital period* of p = 3 years.

> Now that you have finished the experiment,
> use your answers to take the online quiz on the
> *Fundamental College Astronomy Experiments* **ONLINE** website.

EXPERIMENT #5

Classifying Planets

In this experiment you will compare data about objects in our solar system to see why they are classified into different groups.

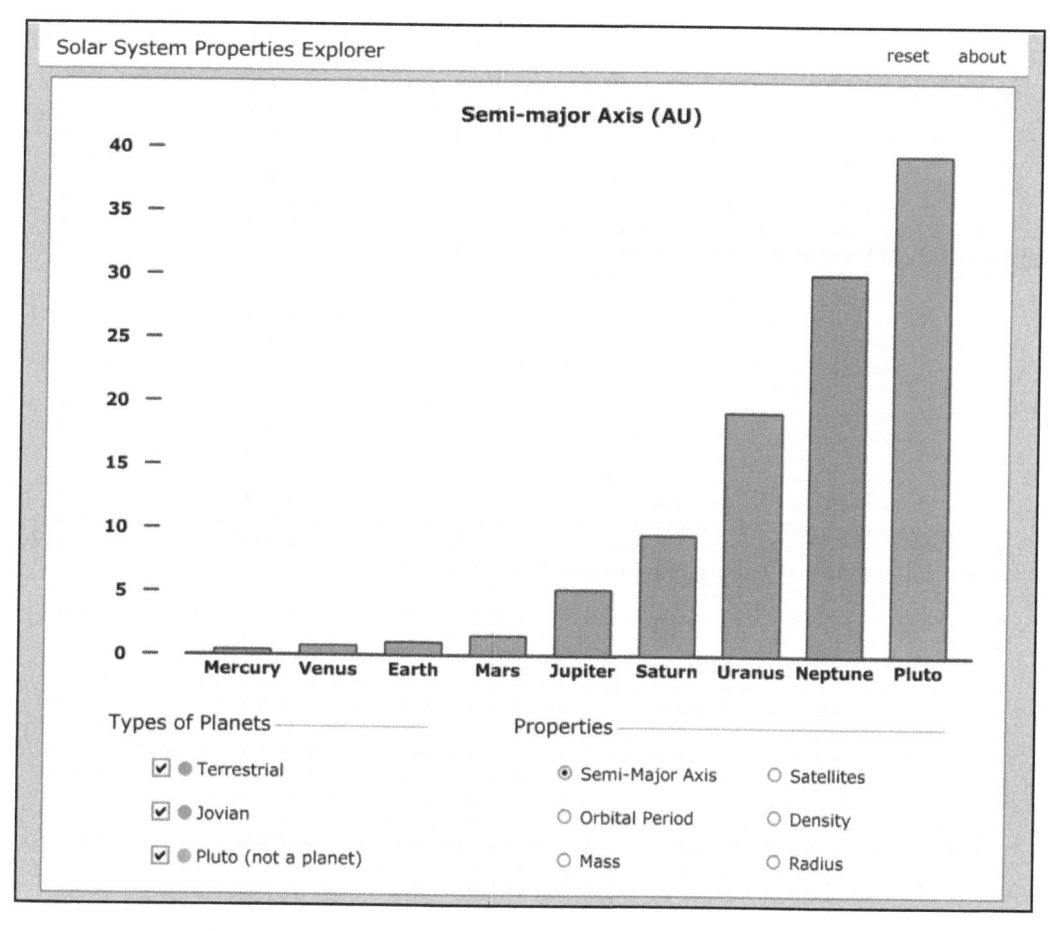

FIGURE 5.1 Solar System Properties Explorer

Simulation downloaded from the Astronomy Education at the University of Nebraska-Lincoln Website (http://astro.unl.edu).

> Use the *Solar System Properties Explorer*, shown in **Figure 5.1**, to answer the questions below.

Which objects are considered Terrestrial Planets?

Which objects are considered Jovian Planets?

Which object is NOT considered a Planet?

Select Semimajor Axis under Properties. How do the distances from the Sun of the Terrestrial Planets compare to those of the Jovian Planets?

Select Orbital Periods under Properties. How do the Orbital Periods around the Sun of the Terrestrial Planets compare to those of the Jovian Planets?

24   Experiment 5   Classifying Planets

Select Mass under Properties. Which Type of Planet is more massive?

Select Radius under Properties. Which Type of Planet is larger?

Select Density under Properties. Which Type of Planet has higher densities?

What can the density tell you about a planet?

Which Terrestrial Planet has a density noticeably lower than the others?

Which Jovian Planet has a density noticeably lower than the others?

To fill out the **TABLE** below, alternately select each of the Properties and place a check mark under the Type of Planet to which Pluto is most similar for that property.

Property	Terrestrial Planets?	Jovian Planets?
Semimajor Axis		
Orbital Period		
Mass		
Radius		
Density		
Satellites		

Does Pluto clearly fit in with either Type of Planet? Explain your answer.

Select Satellites under Properties. Which Type of Planet has more Satellites (moons)?

Can you think of the reason for this?

Which planet(s) has(have) no satellites?

Describe with a SINGLE WORD how all the common Properties of the Terrestrial Planets compare to those of the Jovian Planets.

Now that you have finished the experiment, use your answers to take the online quiz on the *Fundamental College Astronomy Experiments* **ONLINE** website.

26   Experiment 5   Classifying Planets

EXPERIMENT #6

The Transit Method of Extra Solar Planet Detection

In this experiment you will learn about the *Transit Method* for the detection of extra solar planets or exoplanets, planets orbiting stars other than our Sun. A very sensitive light detector, called a photometer, attached to a telescope observes the changes in the amount of light it detects coming from the star. The amount of time the *transit* (or eclipse) takes and by what percentage the detected starlight decreases can tell us the *radius* of a planet and the semimajor axis of its orbit (the distance the planet is from its star).

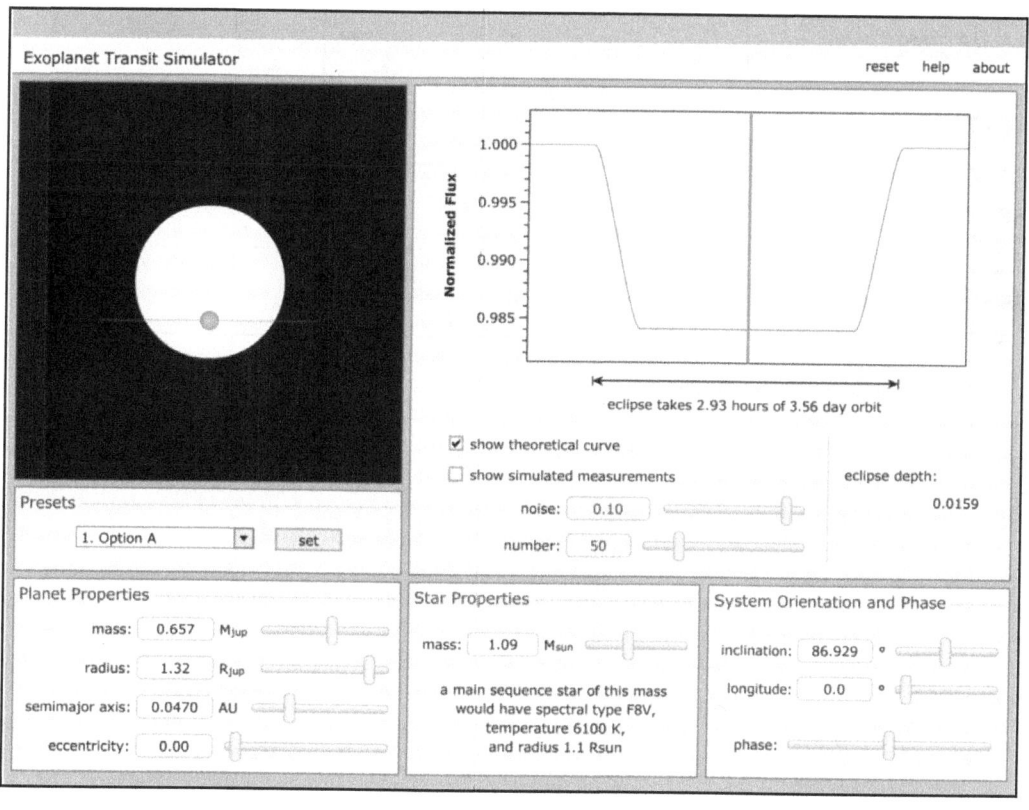

FIGURE 6.1 The Exoplanet Transit Simulator

Simulation downloaded from the Astronomy Education at the University of Nebraska-Lincoln Website (http://astro.unl.edu).

Click on the "help" button for the *Exoplanet Transit Simulator*, shown in the upper right-hand corner of **Figure 6.1**, for instructions on how it works.

First, select Option A under Presets.

Now, use the sliders under Planet Properties to alternately vary the mass and radius of the planet and the semimajor axis (orbital distance) and watch how changing each affects the plot of the transit.

In the **TABLE**, enter whether you observe an *increase*, *decrease*, or *no effect* on **Normalized Flux** (the observed brightness of the star) and **Eclipse (transit) Time** when changing a property. You should press the set button for the Preset Option A after you investigate each change.

Change in Property	Normalized Flux (observed brightness)	Eclipse (transit) time
Increase Mass		
Decrease Mass		
Increase Radius		
Decrease Radius		
Increase Semimajor Axis		
Decrease Semimajor Axis		

Now alternately select each option, 3–11, under Presets and record all the data for each *exoplanet* and its transit in the **TABLE** below. Make sure to click the set button after you select each option. Option 4 is already filled out as an example.

Planet	Mass (Jups)	Radius (Jups)	Semimajor axis (AU)	Normalized Flux	Eclipse (transit) Time (hours)
3.					
4. TrES-1	0.61	1.08	0.0393	0.984	2.62
5.					
6.					
7.					
8.					
9.					
10.					
11.					

Now based on your filled out TABLE above, answer the following questions:

Which exoplanet is _____?

    Most massive —

    Least massive —

    Largest —

    Smallest —

    Orbiting closest to its star —

    Orbiting farthest from its star —

    Decreasing its star's detected light most —

    Decreasing its star's detected light least —

    The one with the longest transit time —

    The one with the shortest transit time —

> Compare the data in your **TABLE** to the data for Presets Option A which is data for *Jupiter* if it were in Earth's orbit and Option B which is Earth.

How do the masses of the exoplanets in your table compare to that of Jupiter and Earth?

How do the radii (sizes) of the exoplanets in your table compare to that of Jupiter and Earth?

How do the semimajor axis (orbital distances) of the exoplanets in your table compare to that of Jupiter and Earth?

How do you think the temperatures on these exoplanets would compare to those of Jupiter and on Earth? Explain your answer.

> Now that you have finished the experiment,
> use your answers to take the online quiz on the
> *Fundamental College Astronomy Experiments* **ONLINE** website.

EXPERIMENT #7

The Radial Velocity Method of Extra Solar Planet Detection

In this experiment you will learn about the *Radial Velocity Method* for the detection of extra solar planets or exoplanets, planets orbiting stars other than our Sun. A telescope is used to observe changes in the wavelength of the light coming from a star that is wobbling due to the gravitational tugs of an unseen planetary companion. The amount of time the starlight takes to go through one complete cycle of these changes depends on the period of the planet's orbit around the star, which allows the semimajor axis of its orbit (the distance the planet is from its star) to be calculated. The amount of wobble and the mass of the star allows the mass of the planet to be calculated.

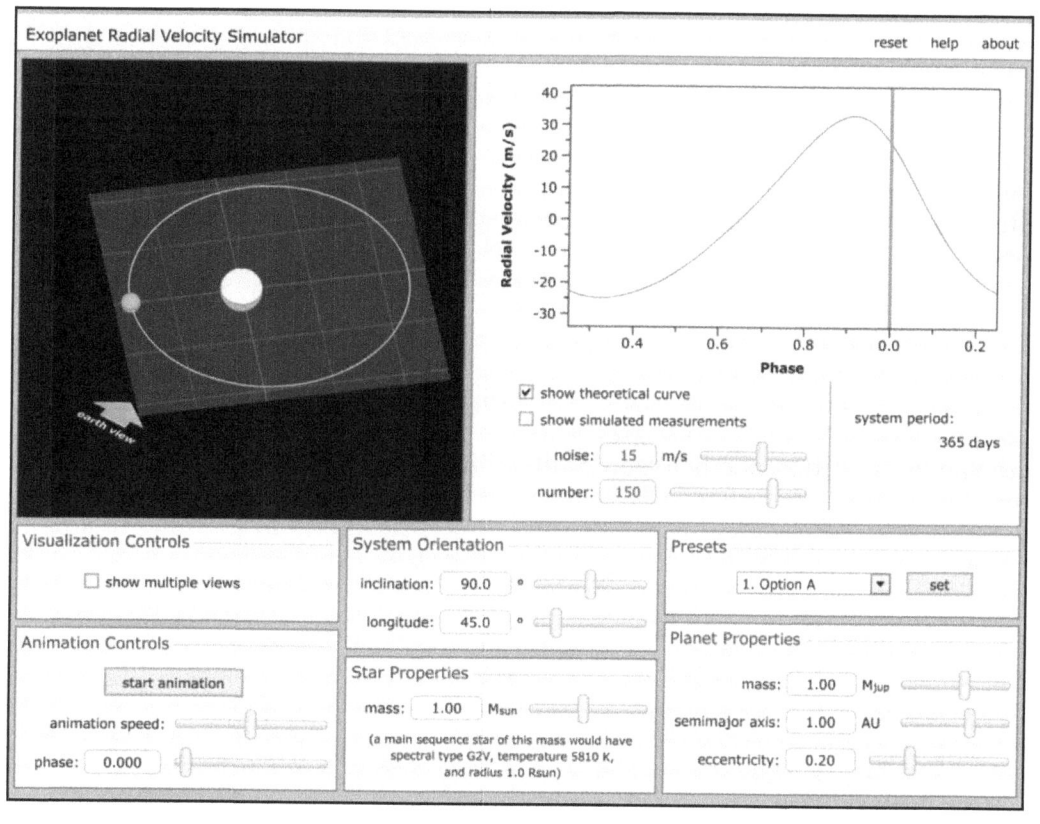

FIGURE 7.1 The Exoplanet Radial Velocity Simulator
Simulation downloaded from the Astronomy Education at the University of Nebraska-Lincoln Website (http://astro.unl.edu).

31

Click on the "help" button for the *Exoplanet Radial Velocity Simulator*, shown in the upper right-hand corner of **Figure 7.1** for instructions on how it works.

First, select Option A under Presets.

Now, use the sliders under Planet Properties to alternately vary the mass and the semimajor axis (orbital distance) and watch how changing each property affects the **Radial Velocity.**

Also adjust the eccentricity of the orbit and note any changes in the plot (the graph) of the **Radial Velocity** and the Visualization of the orbit, the image to the left of the plot.

What effect does increasing mass of the planet have on the **Radial Velocity**?

What effect does decreasing mass of the planet have on the **Radial Velocity**?

What effect does increasing the semimajor axis of the planet's orbit have on the **Radial Velocity**?

What effect does increasing the semimajor axis of the planet's orbit have on the **Radial Velocity**?

What is the apparent difference between orbits of high and low eccentricity?

Now alternately select each option, 5–7, under Presets and record all the data for each *exoplanet* and its orbit in the **TABLE** below. Make sure to click the set button after you select each planet. When recording the **Radial Velocity,** record the maximum number it reaches whether the number is positive or negative.

	Planet Name	**Mass (Jupiter)**	**Semimajor Axis (AU)**	**Eccentricity**	**Radial Velocity m/s**
5					
6					
7					

> Use the TABLE to answer the following questions:

Which planet is most massive?

Which planet is least massive?

Which planet orbits closest to its star?

Which planet orbits farthest from its star?

Which planet's orbit is most eccentric?

Which planet's orbit is least eccentric?

> Now that you have finished the experiment,
> use your answers to take the online quiz on the
> *Fundamental College Astronomy Experiments* **ONLINE** website.

EXPERIMENT #8

Terrestrial Planet Surfaces

The features that make the surface of each terrestrial planet unique arise from the fact that each planet has a unique combination of size and proximity to the Sun (or temperature). In this experiment you will discover how planetary surface features are affected by these two major factors.

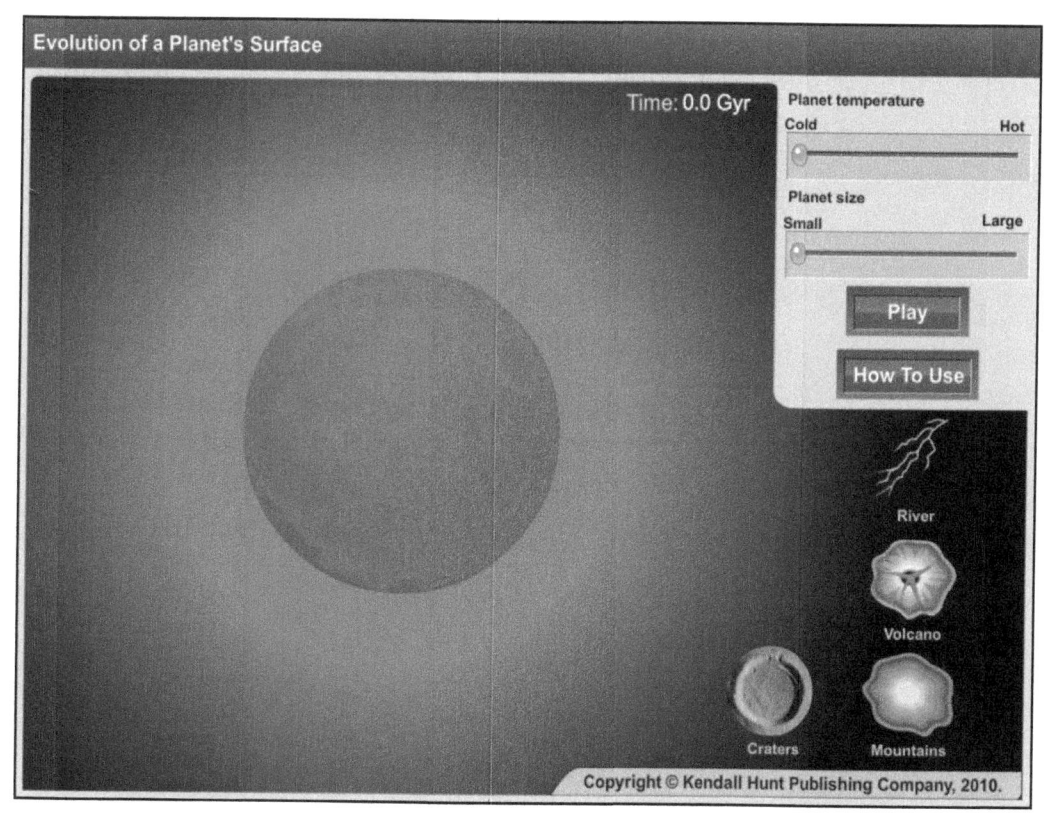

FIGURE 8.1 Evolution of a Planet's Surface Simulator

Use the scroll bars in the upper-right corner of the display, shown in **Figure 8.1**, to alter *Planet temperature* and *Planet size* for a given planet. See the "key" in the lower-right corner of the display for the appearance of each type of surface feature. Click on the "*How To Use*" button for the *Evolution of a Planet's Surface Simulator*, below the "Play" button for instructions.

For each possible combination of temperature and size of a planet in the **TABLE** below, click the "*Play*" button and allow the simulation to run for the full *4.6 Billion Years (GY)*, the current age of the solar system. Circle all features that appear most common at end of each simulation.

Temperature & Size	Cold	Medium	Hot
Small	Craters Rivers Volcanoes Mountains	Craters Rivers Volcanoes Mountains	Craters Rivers Volcanoes Mountains
Medium	Craters Rivers Volcanoes Mountains	Craters Rivers Volcanoes Mountains	Craters Rivers Volcanoes Mountains
Large	Craters Rivers Volcanoes Mountains	Craters Rivers Volcanoes Mountains	Craters Rivers Volcanoes Mountains

> Now based on your completed **TABLE** on the previous page, answer the following questions:

What do all planets with surfaces covered almost completely with craters appear to have in common?

What do planets with rivers appear to have in common?

What do planets with polar ice caps appear to have common?

Planets with little or no water (in the form of rivers or ice caps) appear to be either _____ in size and/or have a _____ surface temperature.

What size and temperature planet appears to be most like Earth?

Figure 8.2 Surface features of our solar system's *terrestrial* planets.
Images courtesy of NASA.

Figure 8.2 above shows the *terrestrial* planets of our solar system in the correct order from left to right of their distances from the Sun (except remember that *Earth's moon* is about the same distance from the Sun as *Earth*) and with their correct relative sizes. Complete the **Surface Temperature** and **Size** columns of the **TABLE** below by selecting the correct relative temperature for each planet based on the distance from the Sun shown in **Figure 8.2** and the relative size of each planet based in the images of the planets in **Figure 8.2**.

Planet	Surface Temperature	Size	Simulation Accurate?
Mercury	Cold / Medium / Hot	Small / Medium / Large	Yes / No
Venus	Cold / Medium / Hot	Small / Medium / Large	Yes / No
Earth	Cold / Medium / Hot	Small / Medium / Large	Yes / No
Earth's Moon	Cold / Medium / Hot	Small / Medium / Large	Yes / No
Mars	Cold / Medium / Hot	Small / Medium / Large	Yes / No

Determine whether or not the *Evolution of a Planet's Surface Simulator* is accurate for each of our solar system's *terrestrial* planets by comparing the pictures and descriptions of each planet's surface in **Figure 8.2** with the simulation's prediction for a planet of its same size and temperature from the first **TABLE**. Circle Yes or No for each planet in the above **TABLE**.

> Now based on **BOTH** of the **TABLES** you completed and inspection of **Figure 8.2**, answer the following questions.

Which two objects in **Figure 8.2** and the second **TABLE** have basically the same surface conditions?

What are the surface conditions of these two objects?

Which factor, surface temperature or size, do these two objects have more in common?

So the similarity in the surface conditions of these two objects is caused by which factor?

Which two objects in **Figure 8.2** and the second **TABLE** are about the same distance from the Sun?

Do these two objects have similar or different surface conditions?

Which factor, surface temperature or size, is different about these two objects?

So the difference in the surface conditions of these two objects is caused by which factor?

Do *Earth* and *Venus* have different or similar surface conditions?

Which factor, surface temperature or size, do these two objects have more in common?

Which factor, surface temperature or size, is different about these two objects?

So which factor causes the comparison you made between their surface conditions?

Most of our solar system's *asteroids* are all smaller than Earth's Moon and are located in a belt that is further from the Sun than Mars.

So the surface temperature of an *asteroid* should be Cold / Medium / Hot (circle one).

The size of an *asteroid* should be Small / Medium / Large (circle one).

Based on what you have learned, which object(s) in **Figure 8.2** do you think should have the most similar surface conditions to an asteroid?

Describe the surface conditions of on asteroid.

40 Experiment 8 Terrestrial Planet Surfaces

Use the first **Table** and/or run the *Evolution of a Planet's Surface Simulator* to test your prediction.

    Was your prediction accurate? Yes / No (circle one).

The moons of the outer *Jovian* planets in our solar system are at most about the size of Mercury or Earth's Moon and all these planets and moon are over 5 times as far from the Sun as Earth.

    So the <u>temperature</u> of most *Jovian* moons should be:

        Cold / Medium / Hot (circle one)

    The <u>size</u> of most *Jovian* moons should be:

        Small / Medium / Large (circle one)

Based on what you have learned, which object(s) in **Figure 8.2** do you think should have the most similar surface conditions to most *Jovian* Moons?

Describe the surface conditions of <u>most</u> *Jovian* Moons.

Use the first **TABLE** and/or run the *Evolution of a Planet's Surface Simulator* to test your prediction.

Was your prediction accurate? Yes / no (circle one).

**Now that you have finished the experiment,
use your answers to take the online quiz on the
Fundamental College Astronomy Experiments ONLINE website.**

EXPERIMENT #9

Telescopes and Atmospheric Absorption

In this experiment you will explore at what altitude above the ground telescopes must be placed in order for astronomers to observe different wavelengths of electromagnetic radiation, different types of light, emitted by astronomical objects.

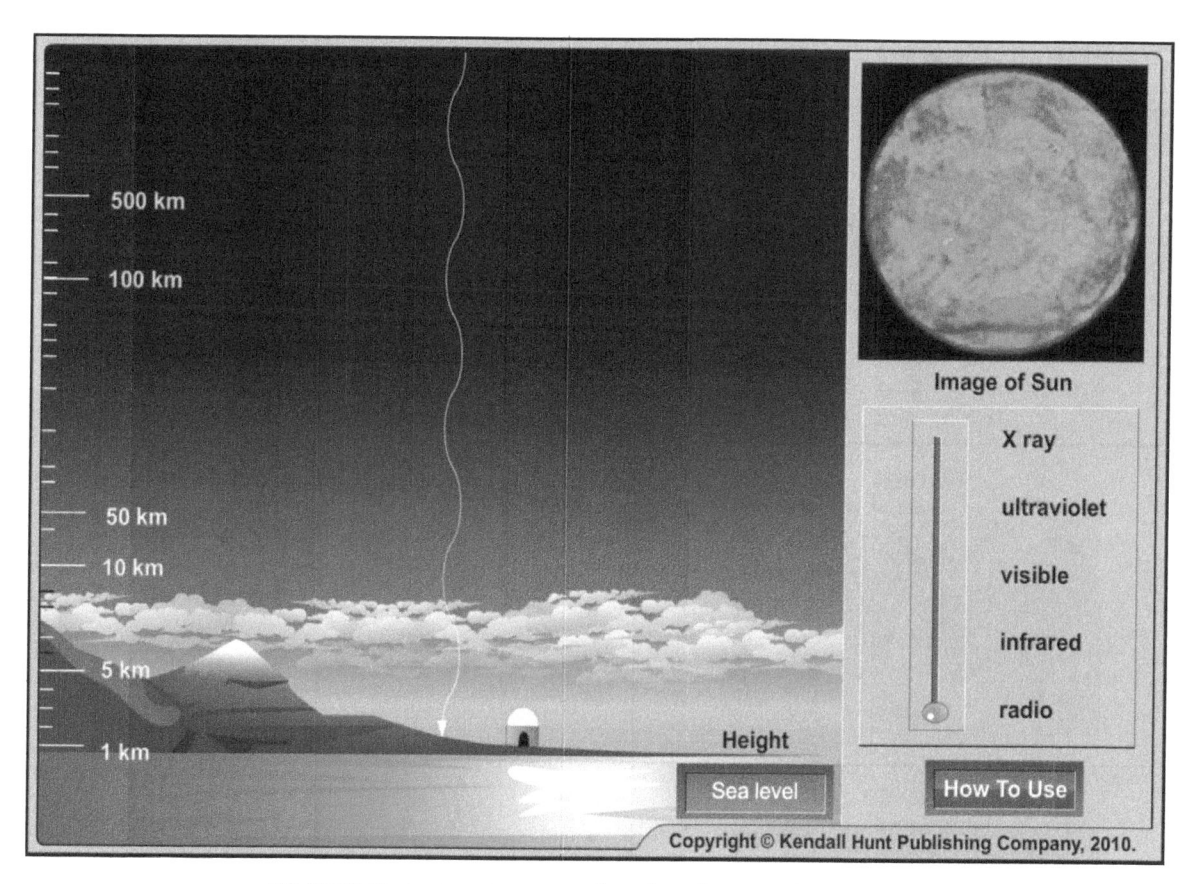

FIGURE 9.1 Atmospheric Absorption Tool

Click on the "*How To Use*" button in the bottom-right corner of the *Atmospheric Absorption Tool,* shown in **Figure 9.1**, for instructions.

Start with the *Atmospheric Absorption Tool,* as you first find it, shown in **Figure 9.1**, set for radio waves. Drag the figure of a telescope upward from Sea Level and record the Height above Sea Level at which telescopes are first placed in each different type of *vehicle* listed in the **TABLE** below. Record the Heights in the **Table**.

Type of Vehicle	Height (km above sea level)
Ground-Based Telescope	0 km (sea level)
Airplane	
Balloon	
Rocket	
Satellite	

Photos © Shutterstock.com

What is the range of Heights at which a telescope must be placed in each type of *vehicle*?

Ground-based

Airplane

Balloon

Rocket

Satellite

How would you place a ground-based telescope at 7 km above sea level?

> **Complete the TABLE on the next page before answering the questions below.**

Which type(s) of light can be observed from telescopes on the ground?

Which types(s) of light require a telescope be placed on an airplane to be observed?

Which type(s) of light requires a telescope be placed on a balloon or rocket to be observed?

Which type(s) of radiation can only be observed by a telescope placed on a satellite in space?

Experiment 9   Telescopes and Atmospheric Absorption   45

Now, alternately move the slider in the *Atmospheric Absorption Tool* to each different type of electromagnetic radiation to observe how deep into the atmosphere each type of electromagnetic (light) wave penetrates. Also note the different wavelengths of the different types of light. Now click on and drag the image of the telescope to raise it to the minimum Height necessary to observe each of the different types. The image of the telescope will change to an image of the necessary vehicle type. Record your results in the **TABLE** below.

Type of EM radiation (light)	Minimum penetration Height above ground	Type of observation vehicle
Radio		
Infrared		
Visible		
Ultraviolet		
X ray		

Use the *Atmospheric Absorption Tool* to list the five (5) types of electromagnetic radiation (light) from the **TABLE** in order of decreasing wavelength, from longest to shortest.

Now that you have finished the experiment,
use your answers to take the online quiz on the
Fundamental College Astronomy Experiments **ONLINE** website.

EXPERIMENT #10

Blackbody Curves

The spectrum of different colors emitted by hot, dense objects is called the **Blackbody Spectrum**. By analyzing an object's blackbody spectrum, we can determine an object's temperature. In this experiment you will explore how blackbody spectra differ between objects at different temperatures.

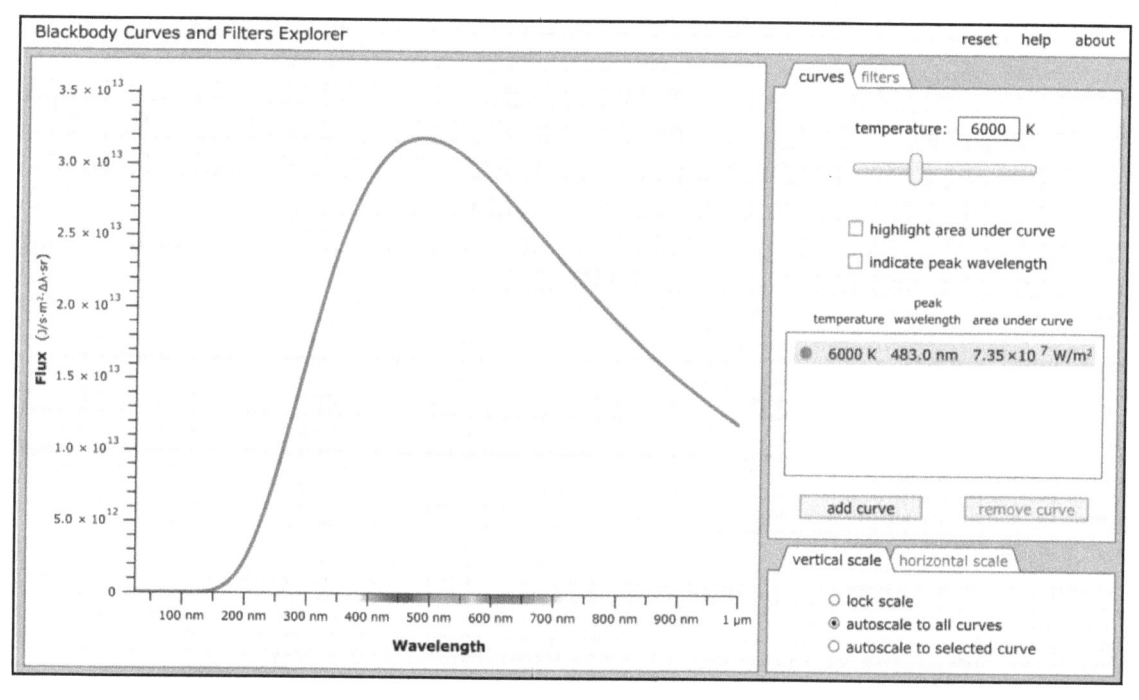

FIGURE 10.1 Blackbody Curves and Filters Explorer
Simulation downloaded from the Astronomy Education at the University of Nebraska-Lincoln Website (http://astro.unl.edu).

Click on the "*help*" button in the top right corner of the *Blackbody Curves and Filters Explorer*, shown in **Figure 10.1**, for instructions.

To start, also click the "*reset*" button in the top right corner to make sure your *Blackbody Curves and Filters Explorer* has the same settings as shown in **Figure 10.1**.

Click on the add curve button to make sure the curve and the information for the star of temperature 6000K is preserved. Also check the indicate peak wavelength box. Now record the peak wavelength and area under curve of the 6000 K star in the **TABLE** below.

Temperature (K)	Peak Wavelength (nm)	Color	Area Under Curve (W/m)^2
7500			
6000			
4500			

Adjust the sliding temperature scale as close to 7500 K as you can and record the peak wavelength and area under the curve. Click the add curve button to make sure the curve and the information is preserved. Do the same thing for a star of temperature 4500 K.

The colors of visible light shown on the **Wavelength** axis tell you the color that it appears to be. Stars with a peak wavelength is approximately above the middle wavelength of the visible spectrum appear yellow, stars with a peak wavelength approximately above the short wavelength end of the visible spectrum appear blue, and those with a peak wavelength approximately above the long wavelength end of the spectrum appear red. Based on the temperature of each star, record its apparent color in the **TABLE** above.

Based on the above **TABLE** you have filled out, answer the following questions.

A star with a longer peak wavelength is HOTTER / COOLER (circle one).

A star with a shorter peak wavelength is HOTTER / COOLER (circle one).

List the three star colors you recorded in order from hottest to coolest.

The **Flux** axis of the plot and area under the curve in W/m^2 (Watts per square meter) you recorded in the **TABLE** tells you how much energy the star emits, or more simply, how bright the star is.

Based on the previous **TABLE**, for stars of a given size, RED / YELLOW / BLUE (circle one) stars are the brightest and RED / YELLOW / BLUE (circle one) are the dimmest. So, HOTTER / COOLER (circle one) stars tend to be brighter and HOTTER / COOLER (circle one) stars tend to be dimmer.

> Now look very CAREFULLY at your **Blackbody Curves** to answer these last questions.

Which temperature star from your **TABLE** gives off the most BLUE light?

Which temperature star from your **TABLE** gives off the most RED light?

Experiment 10   Blackbody Curves

Now that you have finished the experiment,
use your answers to take the online quiz on the
Fundamental College Astronomy Experiments **ONLINE** website.

EXPERIMENT #11

The Hertzsprung–Russel Diagram

The Hertzsprung–Russel (H–R) Diagram is a plot or graph of Luminosity (brightness) vs. the Temperature of stars. It is used by Astronomers to categorize different types of stars. In this experiment you will use the H–R Diagram to learn about the properties of different types of stars.

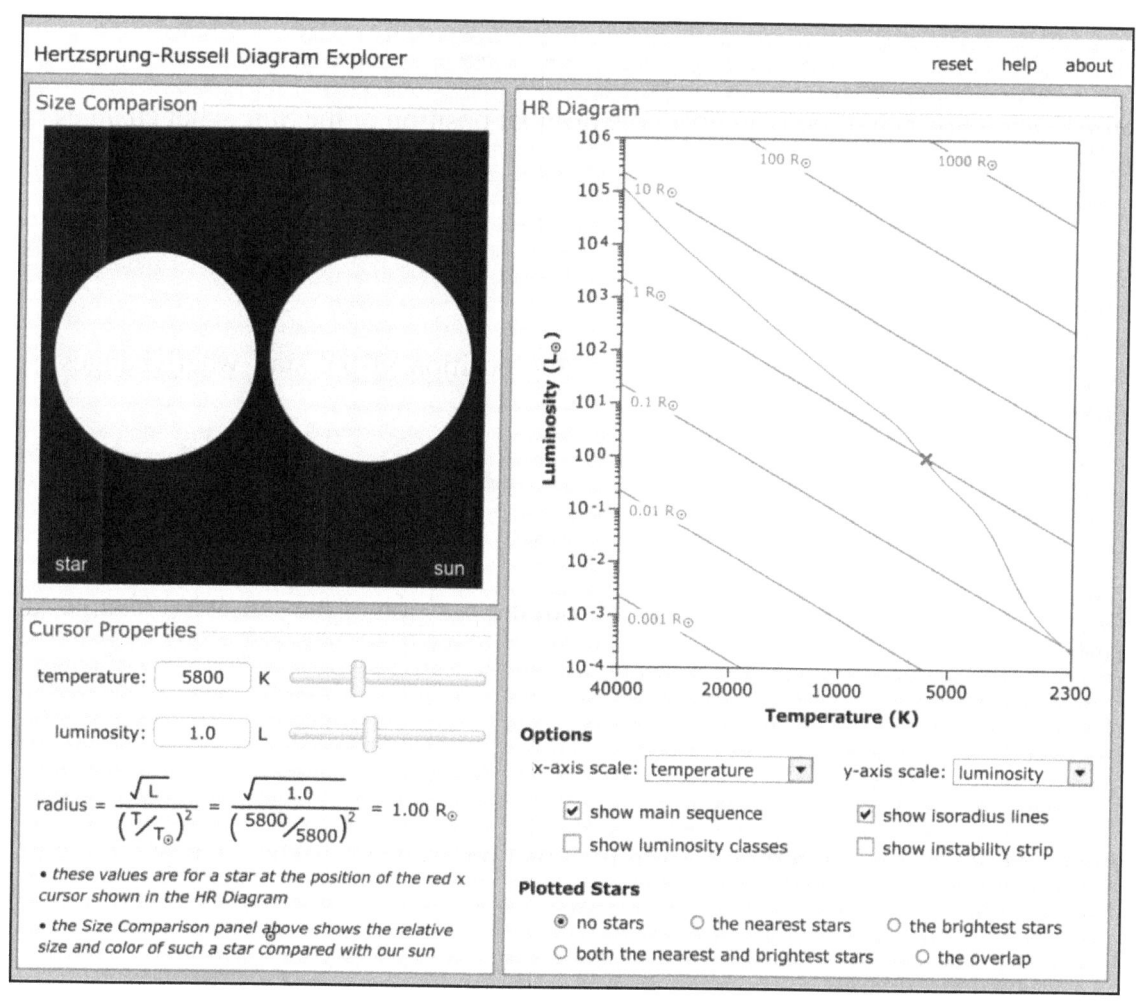

FIGURE 11.1 Hertzsprung–Russel Diagram Explorer
Simulation downloaded from the Astronomy Education at the University of Nebraska-Lincoln Website (http://astro.unl.edu).

Click on the *"help"* button in the top right corner of the *Hertzsprung–Russel Diagram Explorer* shown in **Figure 11.1** for instructions.

To start, also click the *"reset"* button in the top right corner to make sure your *Hertzsprung–Russel Diagram Explorer* has the same settings as shown in **Figure 11.1**.

After the reset, the star in the position of the red "x" on your *H–R diagram* will be our Sun, a *yellow medium-sized* star. All star-types on the *H–R Diagram* can be named in terms of their *color*, which is related to their temperature and their luminosity, which is related to their *size*.

Click on and drag the red "x" around the H–R Diagram. Move it both up and down from the position of our Sun and back and forth from the position of our Sun and watch the changes that occur to the star next to the Sun, the Size Comparison window, and the temperature, luminosity (brightness), and radius in the Cursor Properties window. Do this to answer the questions below.

When you move the star directly up from the original position of the Sun, what changes do you notice in its properties?

When you move the star directly down from the position of the star, what changes do you notice in its properties?

When you move the star directly to the right from the position of the star, what changes do you notice in its properties?

When you move the star directly left from the position of the star, what changes do you notice in its properties?

Of the colors blue, red, and yellow, which is the color of the hottest stars?

Which is the color of the coolest stars?

What other star colors do you notice?

What in general do you notice about the size of brighter stars compared to the size of dimmer stars?

TABLE (see directions BELOW)

Star #	Temperature (K)	Luminosity (Sun = 1)	Radius (Sun = 1)	Color	Size
1	20,000		0.01	White	
2*			0.1		
3 (Sun)	5800	1	1	Yellow	Medium
4*	25,000				
5		2500	100		

An asterisk * indicates that the star is on the *main sequence*, the **red line** on your H–R Diagram.

Use the clues already in the **TABLE** above to complete all the entries. Once you have correctly placed a star by dragging it to the appropriate location on the H–R Diagram, find the temperature, luminosity, and radius in the Cursor Properties window and judge the color and size of a star from the Size Comparison window. Stars that are similar size to the Sun are called *medium* sized, stars that are much bigger are called *giants,* and stars that are much smaller are called *dwarfs*.

Now based on the *color* and *size* of each star in the above **TABLE** that you filled out, name the type of each star in the **TABLE** below.

Star #	Star Type
1	
2	
3 (SUN)	Yellow–Medium
4	
5	

Experiment 11   The Hertzsprung-Russel Diagram

Use your *HR Diagram* to fill out the first three rows of the **TABLE** below by moving your star to the given radius and temperature and recording the luminosity. Use the given temperature and luminosity to record the radius of the star in the last row.

Radius (Sun = 1)	Temperature (K)	Luminosity (Sun = 1)
1	20,000	
1	5000	
10	5000	
	5000	150

Now answer the following questions:

If two stars are the same size, what could cause one to be brighter than the other?

If two stars are the same temperature, what could cause one to be brighter than the other?

If two stars have the same brightness and one is hotter than the other, what must also be true about the hotter star compared to the cooler star?

If two stars have the same brightness and one is smaller than the other, what must also be true about the smaller star compared to the bigger star?

Now that you have finished the experiment,
use your answers to take the online quiz on the
Fundamental College Astronomy Experiments **ONLINE** website.

EXPERIMENT #12

Hubble's Law

Hubble's Law is a relationship between the distance of faraway galaxies from Earth and the velocity with which they are moving away from Earth. In this experiment you will use Hubble's Law to determine the amount of time the galaxies have been moving away from each other to estimate the actual age of the universe.

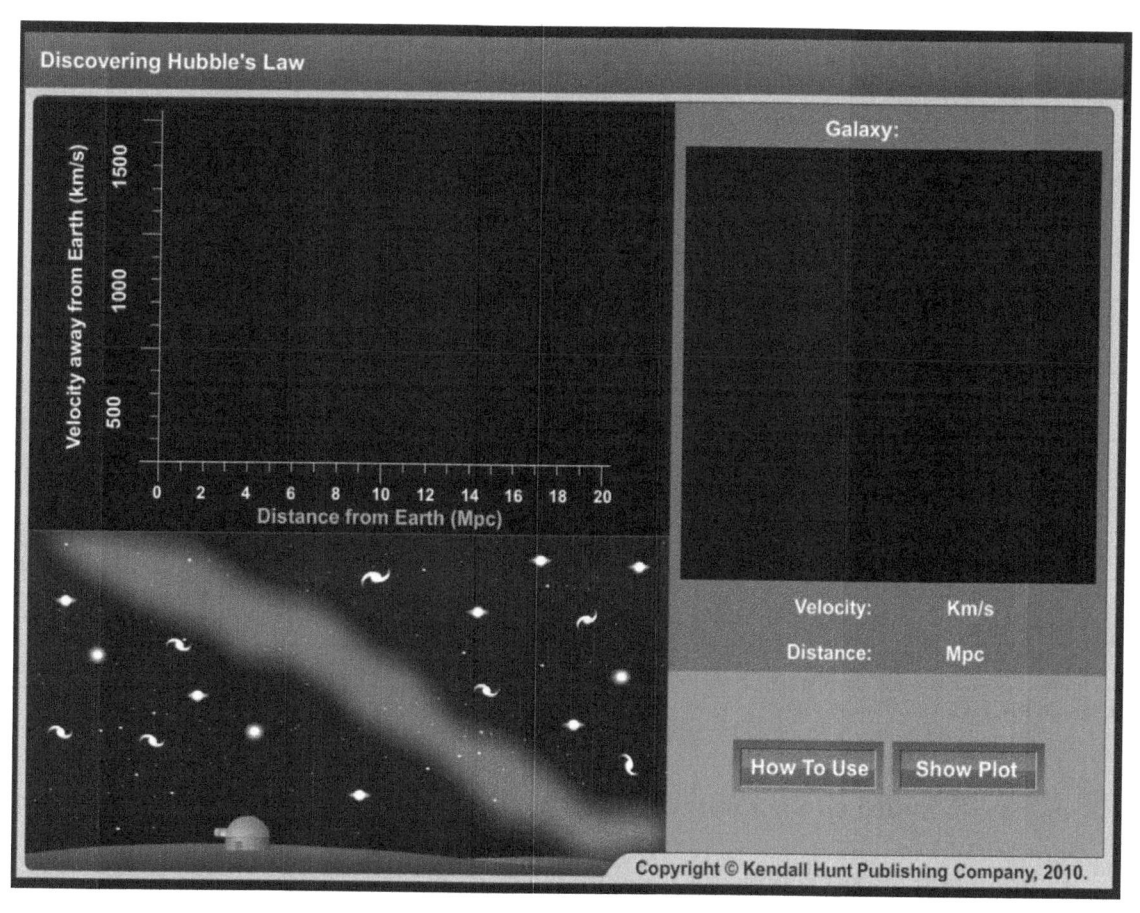

FIGURE 12.1 Discovering Hubble's Law

Click on the "*How To Use*" button in the bottom-right corner of *Discovering Hubble's Law* shown in **Figure 12.1**, for instructions.

Alternately click on each of the 17 galaxies shown in the bottom-left window of *Discovering Hubble's Law* and record the information that appears in the upper-right Galaxy window in the **TABLE** below. You will notice a point appearing on the plot in the upper-left corner each time you click on a galaxy. Some of the information is already recorded for you.

Galaxy	Distance from Earth (Mpc)	Velocity Away from Earth (km/s)
M104		
		920
	4.4 (not 44.4)*	

* If your data read 44.4 it was a mistake; it should be 4.4 which agrees with the numbers on the Distance from Earth (Mpc) axis of the plot.

Once you have recorded all the information in the **TABLE** above, click on "*Show Plot*" in the lower-right window of *Discovering Hubble's Law* to draw a "best-fit" line through the data; this is the straight line that best represents the data.

> Look at the plot, best-fit line, and your data in the TABLE to answer the following questions.

Galaxies that are farther away from Earth are moving away from Earth at a FASTER / SLOWER (circle one) velocity than those that are closer to Earth.

Since all the galaxies are moving away from Earth, what does this suggest about the universe as a whole?

The slope of the best-fit line will be your data's value for *Hubble's Constant*, H.

You will determine the slope of your line by first identifying the TWO POINTS in your data that actually lie ON your best-fit line. Circle both of them in the **TABLE** above.

The slope of a line is a numerical measurement of how steep it is. This is calculated by dividing the RISE of the line over the RUN of the line.

$$SLOPE = RISE / RUN$$

The RISE axis of your graph is the **velocity**, so to calculate the RISE subtract the *smaller* of the TWO velocities you circled in your TABLE from the *larger* one.

$$RISE = v_{large} - v_{small} = \underline{\hspace{1cm}} - \underline{\hspace{1cm}} = \underline{\hspace{1cm}}$$

The RUN axis of your graph is the **distance**, so to calculate the RUN subtract the *smaller* of the TWO distances to the SAME TWO GALAXIES you circled in your TABLE from the *larger* one.

$$RUN = d_{large} - d_{small} = \underline{\hspace{1cm}} - \underline{\hspace{1cm}} = \underline{\hspace{1cm}}$$

Now to determine your SLOPE, divide the RISE by the RUN.

$$SLOPE = RISE / RUN = \underline{\hspace{1cm}}$$

The SLOPE you calculated is your value of *Hubble's Constant* in the units of km/s/Mpc (Mpc = Megaparsecs).

$$H = \underline{\hspace{1cm}} \text{ km/s/ Mpc}$$

Hubble's Law is a relationship between how far away from Earth galaxies are (*distance—d*) and how fast they are moving away from Earth (*velocity—v*). *Velocity* and *distance* are also related by time, $v = d/t$, and since *Hubble's constant* is the slope of your line, it is related to *distance* and *velocity* by $H = v/d$. Now, because $v/d = 1/t$, then $H = 1/t$ or $t = 1/H$. The amount of time that the galaxies have been moving away from Earth is the reciprocal of *Hubble's constant*. This is literally the **age of the universe** calculated with Hubble's constant. Use your value of *Hubble's constant* in the formula below to calculate the **age of the universe**.

Divide your value of *Hubble's constant* into 1000 to calculate the **age of the universe** (the factor of 1000 converts the units to billions of years).

_____ $T = 1000/H =$ _____ billion years

Now that you have finished the experiment,
use your answers to take the online quiz on the
Fundamental College Astronomy Experiments **ONLINE** website.

EXPERIMENT #13

Epilogue–The Drake Equation

Do you think that there might be anyone out there sending or receiving messages with the hope of communicating? The *Drake Equation* was developed as an aid to help answer this question. The Drake Equation attempts to answer the question "How many technically advanced civilizations exist within our own galaxy?" There is no "correct" answer to the question, but the Drake Equation provides a critical thinking path to an answer for you. In this experiment you will come up with your own estimate of the number of advanced life forms that might have developed in our galaxy.

Go to the website:

http://www.activemind.com/Mysterious/Topics/SETI/drake_equation.html

Read the description of the Drake Equation at the site and use the tool at the site to make your own estimate for *N*- the number of communicating civilizations in our galaxy.

> Record *your* estimates (the ones at the site are just examples) for each factor in the equation.

N_* = the number of stars in the Milky Way galaxy = _____

f_p = fraction of stars with planets around them = _____

n_e = number of planets per star ecologically able to sustain life = _____

f_l = fraction of those planets where life actually evolves = _____

f_i = the fraction of f_l that evolve intelligent life = _____

f_c = the fraction of f_i that communicate = _____

f_L = the fraction of the planet's life during which the communicating civilizations survive = _____

Use the tool at the site to make your own estimate for **N**- the number of communicating civilizations in our galaxy.

$$N = N^* f_p n_e f_l f_i f_c f_L$$

N = the number of communicating civilizations in the galaxy

N = _____

Divide your estimate for N*, the number of stars in the Milky Way galaxy. By your final value of N, the number of communicating civilizations in the galaxy:

N*/ N = _____ / _____ = _____

Now divide your answer into one (1)

1/ _____

This tells you, according to your estimates for the Drake Equation, that one out of this many stars in our galaxy has a planet in orbit with a civilization with which we could communicate.

Now, based on your answer:

Other life in the galaxy is probably

common / rare / hard to tell (circle one).

Other civilizations in the galaxy are probably

common / rare / hard to tell (circle one).

Communication with other life/civilizations in the galaxy is probably

likely / unlikely / hard to tell (circle one).

**Now that you have finished the experiment,
use your answers to take the online quiz on the
Fundamental College Astronomy Experiments ONLINE website.**